YOUR KNOWLEDGE HAS VALUE

Manish Abraham

Technology is inspired by culture. Technology assumes new meanings in different cultures

Cultural Integration & collaboration by the use of Technology

GRIN Verlag

Bibliografische Information der Deutschen Nationalbibliothek:

Die Deutsche Bibliothek verzeichnet diese Publikation in der Deutschen National-
bibliografie; detaillierte bibliografische Daten sind im Internet über http://dnb.d-
nb.de/ abrufbar.

Imprint:

Copyright © 2014 GRIN Verlag GmbH
Druck und Bindung: Books on Demand GmbH, Norderstedt Germany
ISBN: 978-3-656-59558-8

This book at GRIN:

http://www.grin.com/en/e-book/268501/technology-is-inspired-by-culture-techno-
logy-assumes-new-meanings-in-different

GRIN - Your knowledge has value

Der GRIN Verlag publiziert seit 1998 wissenschaftliche Arbeiten von Studenten, Hochschullehrern und anderen Akademikern als eBook und gedrucktes Buch. Die Verlagswebsite www.grin.com ist die ideale Plattform zur Veröffentlichung von Hausarbeiten, Abschlussarbeiten, wissenschaftlichen Aufsätzen, Dissertationen und Fachbüchern.

Visit us on the internet:

http://www.grin.com/

http://www.facebook.com/grincom

http://www.twitter.com/grin_com

University of Malta

Technologies are imbued (inspired) by values pertinent to a particular culture. A look at cases of technologies that have been integrated into a culture and a look into the way the technology assumes new meanings.

An additional insight into the cultural integration & collaboration by the use of technology.

Author - Manish Abraham
Edward de Bono Institute, University of Malta, Msida, Malta,

In this paper there is an effort made to show how technology gets inspired by values pertaining to a certain culture and there is also an attempt made to show various scenarios, where technology is integrated into a culture and look at how culture & social situations define new definitions to these kind of technologies further on there is an attempt made to show the interrelation and how technology improves cultural integration and progress.

Today there are around 7 billion people in this world[1] and around 6 billion of them have access to a mobile phone[2] and around 3 billion have access to the Internet[3], it is used in different ways in different scenario, situations, and uses, what this shows is that there is a very diverse fascinating and interesting intersection in its use.

People normally react to new technology in a very strange way and say that there is something wrong, but in due course of time people start adapting that technology, and then technology starts getting inspired by the way a person, community or culture thinks (Rip, 1995).

Initially, when Walkman was launched we saw that many people argued that it would make an individual away from the social world and give them time to concentrate on his own pleasure, which would make him or her away from the society. Later we observe that people adapted a Walkman (Gay, 1997) and further new solutions like I pod & mp3 players were made.

---

[1] Population Clock. 2014. Available at: http://www.census.gov/popclock/.
[2] BBC News - UN: Six billion mobile phone subscriptions in the world. 2014. Available at: http://www.bbc.co.uk/news/technology-19925506.
[3] World Internet Users Statistics Usage and World Population Stats. 2014. Available at: http://www.internetworldstats.com/stats.htm.

In case of mobile phones initially mobile phones were designed for an industrial purpose in a factory by Motorola, the main reason why mobile phones became famous was when Japanese companies were influenced and inspired by their own culture of talking and sharing knowledge with each other.[4]

Technology sometimes gets integrated with a culture and community, in cases when technology cannot integrate, technology adapts to the culture as seen in Kosher phone which was made to adapt to the Jewish culture (which prohibited cameras, text messaging service).[5]

Technology also brings about a change in the laws of a country not only in bringing new laws, but also amending older ones. For example, the land right Bill in America, gave right to the landowner the land below and the air above, which was a 200 year long law, but in 1940s with the coming of airplanes, they had to amend this law in 1946.[6]

Today we observe that in the case of technology there are many stakeholders, who play a major role in its growth that is public, scientists, politicians and media (Vidgen, 1997, pp. 21-46). Culture also influences the meaning of a technology and many times technologies are inspired by the values imbibed in that culture.

Technology, for an example, mobile phones find a very unique integration with culture and the work that an individual does (Palen, 2001, pp. 109-122), today we see that technologies like mobile phones are being used in various roles , in one culture we see that it is used from the simple function

[4] The Mobile Phone: A History in Pictures -CIO.com. (n.d.). *CIO*
http://www.cio.com/article/504135/The_Mobile_Phone_A_History_in_Pictures
[5] http://en.wikipedia.org/wiki/Mobile_phone#Kosher_phones
[6] https://supreme.justia.com/cases/federal/us/328/256/case.html

as a torch light to more complex roles like fishermen using it for marketing and weather reports (Mittal, 2010). In another we observe how it changes the way people think and brings about a change in society (Contarello, 2007, pp. 149-163), we also observed that it has also seen applications in businesses like new apps (Butler, 2011), our daily activities. Technology is slowly getting locked with our culture, and becomes a daily necessity in some other cultures.

In some cultures (African and South Asian), they have a tradition in which people meet and discuss the various factors like visions, dreams and things which influence their day-to-day activities like economics, social and political factors (Zamora, 1990), but in the end of the meetings generally, it was observed that people come up with ideas, but they did not see how these ideas could be implemented.

How could technology facilitate in changing such a dynamic?

They took mobile phone cameras and documented various stories, traditions and cultural aspects that they were experiencing collectively, later they collectively watch these videos to inspire each others, slowly this use of the technology got spread and people started using the mobile phone cameras to document abuses and by this, they collectively rallied around issues which were documented by the use of such a technology which in social sciences is referred as "collectively call for collective action from below" (Olson, 1965). In this case technology found a new meaning, but to do this you need to trust communities, cultures & traditions.

When this is explained in a logical manner, it explains how a social network works (Boyd, 2010, pp. 16-31), but this pondered a question in my mind.

How is such a technology created what are the algorithms behind the databases, the assumptions in the codes that make the technology possible?

Most of these technologies come in a structural way of mapping the world or in other words, the logics like a parent-child relationship, we can observe this in any of the databases, programs or knowledge management systems. Now, since we can see that there are around 6 billion people who use this kind of a technology of mobile phones with very diverse perspectives. We have to collectively rethink the underlying codes that run the technology (Ash, 2013), so we can observe that it is not only technology that shapes culture, but culture also shapes technology.

The way technology advances is dependent on the way a culture or society wants it. This calls in for a continuous improvement of the technology by collectively improving the thoughts behind the technology, normally it is seen that such a technological change is beyond a cultural or social influence and is seen as "technological imperative" (Pacey, 2013), but these kind of models also then deny the possibility of directing technology, consequent to cultural or social change. *"The main argument tends to be its inadequacy and its ideological function in mystifying and furthering the interests certain groups who benefit from technology could change"* (Williams, 1996). But it has been observed during a period of time that different people add to their cultures, ideas and feelings to improve the technology (Williams, 1996) the speed of advancement of technology increases. For example, the android, which itself is an open source platform where advancements takes place from the individual's perspective, people tend to use the basic technology as a foundation and build over it, to get it adapted in their cultural perspective, few examples of this would be applications which allow farmers to switch on the watering machine to applications such as Skype, which was started from the basic concept of collective sharing of information and got its impact into the various segments of the society like businesses, education, travel, research and many more. It has been observed that there is "no inherent or compelling logic of technical development" (Rohracher, 1998). *"Social construction of technology approach (SCOT)"* (Pinch, 1984) gives an insight on how culture and sociology of scientific knowledge helps in technological development.

Vision plays a major role in determining how a technology can grow or be used. Every culture has something very unique and different from the other culture and hence a very different approach to technology, in the olden times, people were not connected and hence what used to be a vision in one culture used to be a technological marvel in the other culture, this can be seen in the case of airplanes. In the Indian culture the concept of a flying machine has been a tale for thousands of years (Childress, 1991) and it was not until Leonardo da Vinci drew a few drawings inspired by pigeons the idea came into the notice to a different culture, in the end credit goes to the developers and the researchers, who developed the aeroplane (Childress, 1991).

In other scenarios it can be a concept. The earth used to be flat in the mediaeval times, for one culture (Russell, 2007) and the earth used to be round revolving around the sun in other and the earth used to be round with some revolving around the earth in some other culture (Diakidoy, 2001). There used to be a boundary of communication, but today some of the common technologies like communication has got deeply integrated with most of the cultures and this gives a chance for us to share our ideas and thoughts that have been prevailing for centuries. This could add to a concept which I developed which gives importance to a collective conscious C (Abraham, 2014), but here I would call it **cultural integration and collaboration by using technology**.

Since the topic did not specify culture to be society bounded, there is also an attempt made to look into various other cultures.

There are many technologies like computers, mobile phones iPad and many more which have also influenced learning culture (Marsick, 2003) let it be an organization or students. There was a time when people used to write and read the archives, which used to be the main source of information (Duff, 2002) but today we see that technology has changed the learning culture, this kind of change has helped to learn stuff more easily and to gather Data more easily, from a different perspective, many also argue this is affecting the thinking capability of the learner as they rely on technological equipment to do things (Koeber, 2005).

Technologies in software, hardware, bionics, and various other fields have influenced cultures in different job profiles, starting from writers to artists to singers. Technologies have changed and integrated with culture of business, journalism, medicine and other fields.

We used to send letters which used to be written on paper, and today most of us send emails, which shows a change in organizational culture. Genetic engineering has come up with solutions, which has helped solve food crisis in many places.[7]

Technology is slowly getting integrated into various cultures. For example, in the ancient culture, people used to light their houses by using candles, or reflecting mirrors (Schivelbusch, 1995) but with the advent of technology like a fluorescent bulb we see that initially it was spread to some section of the society, but in a period of time, many cultures have got used to it and integrated the use of bulbs for light. The basic technology would have advanced from fluorescent to LED, this can even be seen in the source of energy from coal to diesel to Hydro to other nonrenewable sources of energy, but the basic technology has integrated into many cultures and society.

In most of the cultures riding a horse used to be a sign of pride, tradition, ease of transport (Clutton-Brock, 1992) but with the advent of the motor engine slowly we see how there is a progress in the integration of technology.

In future, the role of technology would not only be to replicate the needs of a culture, but also choose parts from various cultures, integrate them and create a new model of foundation for a future technology.

In some of the cultures meeting a person in person was considered to be the right way of communicating, but we observe that nowadays technology has integrated and slowly, people have started adapting different other means of communication like mobile phones, video conferencing etc.

---

[7] http://www.globalissues.org/article/190/ge-technologies-will-solve-world-hunger

Another example of technology getting integrated with a culture would be washing machines, in some cultures, washing clothes used to be a way in which communities used to get together and wash their clothes and today slowly they have adapted to washing machines.

Before gas or induction based stoves, cooking used to be done on fire from wood or coal and there were various rituals, which used to be part of the culture[8] since fire was considered holy in that culture, but slowly in due course of time people started getting adapted to the change.

Culturally, we observe that people used to put salt on meat or dry the meat so that it stays for a long time, but after a refrigerator was invented it took some time till people integrated it into their lifestyle and now we can see a large section of the cultures around the world have got an integrated simple concept of using a refrigerator.

In some cultures (Indian), there was a concept of conscious human existence[9] prevalent for a long time which gave rise to the concept of virtual realities and having different ways of living life which inspired technology which allows people to live a virtual life in games like second life (Kantonen, 2010) this is another example how technology is inspired by a culture.

Technology itself is bringing about collaboration in various cultures by increasing the knowledge about other culture and slowly gets integrated and accepted by most of the cultures. The future scenario does not only depend on technology being adopted by cultures, but also vice versa where technology adapts and learns from culture.

---

[8] http://www.myresources.com.au/downtime/food-and-drink/13-general/5221-cooking-with-fire
[9] http://www.belurmath.org/vedantaindianculture.htm

References

Abraham (2014). Critical review of the 4 C model of creativity and the need for a creative conscious C. University of Malta.

Ash, J. (2013). Rethinking affective atmospheres: Technology, perturbation and space times of the non-human. Geoforum, 49, 20-28.

Boyd, D., & Ellison, N. (2010). Social Network Sites: Definition, History, And Scholarship. IEEE Engineering Management Review, 38 (3), 16-31.

Butler, M. (2011). Android: Changing the mobile landscape. Pervasive Computing, IEEE, 10(1), 4-7.

Childress, D. H. (1991). *Vimana aircraft of ancient India & Atlantis*. Adventures Unlimited Press.

Clutton-Brock, J. (1992). Horsepower: a history of the horse and the donkey in human societies. Natural History Museum Publications.

Contarello, A., Fortunati, L., & Sarrica, M. (2007). Social Thinking And The Mobile Phone: A Study Of Social Change With The Diffusion Of Mobile Phones, Using A Social Representations Framework. Continuum, 21 (2), 149-163.

Diakidoy, I. A. N., & Kendeou, P. (2001). Facilitating conceptual change in astronomy: A comparison of the effectiveness of two instructional approaches. Learning and Instruction, 11(1), 1-20.

Du Gay, P. (Ed.). (1997). *Doing cultural studies: The story of the Sony Walkman* (Vol. 1). Sage.

Duff, W. M., & Johnson, C. A. (2002). Accidentally found on purpose: information-seeking behavior of historians in archives. The Library Quarterly, 472-496.

Jørgensen, M. S., Jørgensen, U., & Clausen, C. (2009). The social shaping approach to technology foresight. Futures, 41(2), 80-86.

Kantonen, T., Woodward, C., & Katz, N. (2010, March). Mixed reality in virtual world teleconferencing. In Virtual Reality Conference (VR), 2010 IEEE (pp. 179-182). IEEE.

Koeber, C. (2005). Introducing multimedia presentations and a course website to an introductory sociology course: How technology affects student perceptions of teaching effectiveness. Teaching Sociology, 33 (3), 285-300.

Marsick, V. J., & Watkins, K. E. (2003). Demonstrating the value of an organization's learning culture: the dimensions of the learning organization questionnaire. Advances in developing human resources, 5(2), 132-151.

Mittal, S., Gandhi, S., & Tripathi, G. (2010). Socio-economic impact of mobile phones on Indian agriculture. New Delhi: Indian Council for Research on International Economic Relations.

Olson, M. (1965). The logic of collective action: public goods and the theory of groups. Cambridge, Mass.: Harvard University Press.

Pacey, A. (2013). Technology: Practice and Culture. Ethics and Emerging Technologies, 27.

Palen, L., Salzman, M., & Youngs, E. (2001). Discovery And Integration Of Mobile Communications In Everyday Life. Personal and Ubiquitous Computing, 5 (2), 109-122.

Pinch, T.J. & Bijker, W. (1984), 'The Social Construction of Facts and Artefacts: or How the Sociology of Science and the Sociology of Technology Might Benefit Each Other', *Social Studies of Science*, **14**, pp. 399-444.

Pinch, T.J. & Bijker, W. (1986), 'Science, Relativism and the New Sociology of Technology: Reply to Russell', Social Studies of Science, 16, (2), pp. 347-360.

Rip, A., Misa, T. J., & Schot, J. (1995). *Managing technology in society*. Pinter Publishers.

Rohracher, H. (1998), Technology Policy: Would Social Studies of Technology Make a Difference?, Heft 27, Graz: IFZ

Russell, J. B. (2007). The Myth of the Flat Earth. inglés). American Scientific Affiliation. Consultado el, 14-03.

Schivelbusch, W. (1995). Disenchanted night: The industrialization of light in the nineteenth century. University of California Pr.

Vidgen, R. (1997). Stakeholders, Soft Systems And Technology: Separation And Mediation In The Analysis Of Information System Requirements. Information Systems Journal, 7(1), 21-46..

Williams, R., & Edge, D. (1996). The social shaping of technology. Research policy, 25(6), 865-899.

Zamora, M. D. (1990). The Panchayat tradition: a north Indian village council in transition 1947-1962. Reliance Publishing House.